A MODEL TO
ASSESS STAFFING NEEDS
IN NUCLEAR MEDICINE

The following States are Members of the International Atomic Energy Agency:

AFGHANISTAN
ALBANIA
ALGERIA
ANGOLA
ANTIGUA AND BARBUDA
ARGENTINA
ARMENIA
AUSTRALIA
AUSTRIA
AZERBAIJAN
BAHAMAS
BAHRAIN
BANGLADESH
BARBADOS
BELARUS
BELGIUM
BELIZE
BENIN
BOLIVIA, PLURINATIONAL
   STATE OF
BOSNIA AND HERZEGOVINA
BOTSWANA
BRAZIL
BRUNEI DARUSSALAM
BULGARIA
BURKINA FASO
BURUNDI
CAMBODIA
CAMEROON
CANADA
CENTRAL AFRICAN
   REPUBLIC
CHAD
CHILE
CHINA
COLOMBIA
COMOROS
CONGO
COSTA RICA
CÔTE D'IVOIRE
CROATIA
CUBA
CYPRUS
CZECH REPUBLIC
DEMOCRATIC REPUBLIC
   OF THE CONGO
DENMARK
DJIBOUTI
DOMINICA
DOMINICAN REPUBLIC
ECUADOR
EGYPT
EL SALVADOR
ERITREA
ESTONIA
ESWATINI
ETHIOPIA
FIJI
FINLAND
FRANCE
GABON
GEORGIA

GERMANY
GHANA
GREECE
GRENADA
GUATEMALA
GUYANA
HAITI
HOLY SEE
HONDURAS
HUNGARY
ICELAND
INDIA
INDONESIA
IRAN, ISLAMIC REPUBLIC OF
IRAQ
IRELAND
ISRAEL
ITALY
JAMAICA
JAPAN
JORDAN
KAZAKHSTAN
KENYA
KOREA, REPUBLIC OF
KUWAIT
KYRGYZSTAN
LAO PEOPLE'S DEMOCRATIC
   REPUBLIC
LATVIA
LEBANON
LESOTHO
LIBERIA
LIBYA
LIECHTENSTEIN
LITHUANIA
LUXEMBOURG
MADAGASCAR
MALAWI
MALAYSIA
MALI
MALTA
MARSHALL ISLANDS
MAURITANIA
MAURITIUS
MEXICO
MONACO
MONGOLIA
MONTENEGRO
MOROCCO
MOZAMBIQUE
MYANMAR
NAMIBIA
NEPAL
NETHERLANDS
NEW ZEALAND
NICARAGUA
NIGER
NIGERIA
NORTH MACEDONIA
NORWAY
OMAN
PAKISTAN

PALAU
PANAMA
PAPUA NEW GUINEA
PARAGUAY
PERU
PHILIPPINES
POLAND
PORTUGAL
QATAR
REPUBLIC OF MOLDOVA
ROMANIA
RUSSIAN FEDERATION
RWANDA
SAINT KITTS AND NEVIS
SAINT LUCIA
SAINT VINCENT AND
   THE GRENADINES
SAMOA
SAN MARINO
SAUDI ARABIA
SENEGAL
SERBIA
SEYCHELLES
SIERRA LEONE
SINGAPORE
SLOVAKIA
SLOVENIA
SOUTH AFRICA
SPAIN
SRI LANKA
SUDAN
SWEDEN
SWITZERLAND
SYRIAN ARAB REPUBLIC
TAJIKISTAN
THAILAND
TOGO
TONGA
TRINIDAD AND TOBAGO
TUNISIA
TÜRKİYE
TURKMENISTAN
UGANDA
UKRAINE
UNITED ARAB EMIRATES
UNITED KINGDOM OF
   GREAT BRITAIN AND
   NORTHERN IRELAND
UNITED REPUBLIC
   OF TANZANIA
UNITED STATES OF AMERICA
URUGUAY
UZBEKISTAN
VANUATU
VENEZUELA, BOLIVARIAN
   REPUBLIC OF
VIET NAM
YEMEN
ZAMBIA
ZIMBABWE

The Agency's Statute was approved on 23 October 1956 by the Conference on the Statute of the IAEA held at United Nations Headquarters, New York; it entered into force on 29 July 1957. The Headquarters of the Agency are situated in Vienna. Its principal objective is "to accelerate and enlarge the contribution of atomic energy to peace, health and prosperity throughout the world".

IAEA HUMAN HEALTH REPORTS No. 19

A MODEL TO
ASSESS STAFFING NEEDS
IN NUCLEAR MEDICINE

INTERNATIONAL ATOMIC ENERGY AGENCY
VIENNA, 2022

COPYRIGHT NOTICE

© IAEA, 2022

Printed by the IAEA in Austria
October 2022
STI/PUB/1965

IAEA Library Cataloguing in Publication Data

Names: International Atomic Energy Agency.
Title: A model to assess staffing needs in nuclear medicine / International Atomic Energy Agency.
Description: Vienna : International Atomic Energy Agency, 2022. | Series: IAEA human health reports, ISSN 2074–7667 ; no. 19 | Includes bibliographical references.
Identifiers: IAEAL 22-01503 | ISBN 978–92–0–131321–8 (paperback : alk. paper) | ISBN 978–92–0–131421–5 (pdf) | ISBN 978–92–0–131521–2 (epub)
Subjects: LCSH: Nuclear medicine — Professional staff. | Nuclear medicine — Management. | Personnel management. | Nuclear medicine physicians.
Classification: UDC 615.849 | STI/PUB/1965

FOREWORD

Staffing costs account for a significant share of any nuclear medicine service budget. The introduction of new imaging modalities and therapeutic procedures may require a reassessment of staffing needs to ensure the most efficient use of resources. Adequate staffing of any health care service is necessary to provide optimal patient care and maintain safety levels and operational efficiency. Up to now, there has not been a model available to assist with the assessment of staffing needs in nuclear medicine departments of different levels of complexity. Without such a model or a standard to apply, assessing staffing requirements in complex nuclear medicine departments can be challenging.

This publication therefore presents a staffing model and tool for nuclear medicine departments and describes the methodology adopted to reflect operational complexities. Calculations from the model are based on the roles and responsibilities of the professionals involved, which include nuclear medicine physicians, nuclear medicine technologists, radiopharmacists, nurses, cyclotron operators and administrative staff. A staffing model for medical physicists in a diagnostic imaging facility is provided in the publication IAEA Human Health Reports No. 15, Medical Physics Staffing Needs in Diagnostic Imaging and Radionuclide Therapy: An Activity Based Approach. Therefore, this publication does not address that specific issue.

The model, which has been created by experts in the field of nuclear medicine, takes into account clinical workload and infrastructure parameters such as quantity and type of equipment, operational level of the radiopharmacy and presence of an in-house cyclotron. In addition to clinical work, which also includes attendance at multidisciplinary meetings and discussions with referring physicians, the model considers staff time spent setting up equipment, implementing quality assurance and quality control procedures, performing user operated maintenance, undertaking continuous professional development, and carrying out administrative tasks.

Before its release, the model was tested in departments of different levels of complexity, which resulted in adjustments to the calculation factors. The model was also updated to reflect the workload of radiopharmacists and radiochemists by applying the staffing specifications provided in the IAEA publication Operational Guidance on Hospital Radiopharmacy: A Safe and Effective Approach.

The IAEA officers responsible for this publication were D. Paez and M. Dondi of the Division of Human Health.

CONTENTS

1. INTRODUCTION

This publication accompanies the 'IAEA tool to assess staffing needs in nuclear medicine' [1], an on-line form based on an Excel model developed by the IAEA. The tool is hosted on the IAEA Human Health Campus, an educational resource website for health professionals [2], and the IAEA's International Research Integration System (IRIS) secure software platform [3]. The publication addresses staffing requirements for the optimal and safe delivery of nuclear medicine services (NMSs), taking into consideration current standards of practice and relevant IAEA publications [4–12], including IAEA Safety Standards Series Nos GSR Part 2, Leadership and Management for Safety [13], and GSR Part 3, Radiation Protection and Safety of Radiation Sources: International Basic Safety Standards [14].

1.1. BACKGROUND

Human resources are one of the most expensive components in any nuclear medicine department. There are no known standards or models that can help to assess the staffing levels required for a small, medium or large department. Changes that affect staffing requirements, such as new grades or classification levels, shifts in roles and responsibilities, and new or changing modalities (e.g. single photon emission computed tomography (SPECT) and positron emission tomography (PET)), may require adjustments to staff deployment practices or policies to ensure the most efficient and effective use of resources. Without regular staffing reviews, such changes can have profound consequences on departmental efficiency, operational safety and financial performance.

1.2. OBJECTIVE

This publication presents a model to calculate staffing needs in terms of full-time equivalents (FTEs) of nuclear medicine physicians, radiopharmacists, nuclear medicine technologists and nurses. The model could be applied in the planning of new departments, prior to the introduction of new technologies, or for periodical reviews of resource utilization. It is intended for use by hospital administrators, department heads and all nuclear medicine practitioners, and addresses the following areas.

1.2.1. Determining adequate staffing levels

To provide adequate and sustainable services, it is essential for a nuclear medicine department to be able to determine adequate staffing levels. Furthermore, this assessment of staffing levels will facilitate budget planning and allocation, ensure an appropriate distribution of personnel, and guarantee the safety of both patients and staff.

1.2.2. Determining optimal staff deployment

The appropriate staffing of an NMS depends on the department's size (small, medium or large) and type (university or non-university based). Therefore, as needs differ from one department to another, the application of a single model for staff deployment is not appropriate. The model presented in this publication attempts as far as possible to take the different situations into account.

1.2.3. Justifying needs

An independent standardized staffing assessment helps to evaluate current and projected needs.

1.2.4. Assessing system risks and identifying quality improvements

A staffing assessment identifies the risks that may arise from staff shortages and provides clarity on the improvements necessary in terms of quality management.

1.2.5. Improving personnel effectiveness

A staffing assessment may identify an inefficiency or an area of weakness that can be improved through additional staff training or through the introduction of new technology, thereby freeing up staff for more critical responsibilities or post assignments. The assessment also provides a framework for the intelligent planning of posts. By analysing multiple factors (e.g. current staffing levels or normal operating practices), post planning helps to define acceptable levels of overtime and identify available resources or other possible remedies.

Guidance provided here, describing good practices, represents expert opinion but does not constitute recommendations made on the basis of a consensus of Member States.

1.3. SCOPE

This publication describes the assumptions on which the IAEA tool to assess staffing needs in nuclear medicine [1] is built. The underlying reasons and assumptions, based on international guidelines and agreement among the experts involved in this publication, are explained.

Staffing needs are linked to workload and, to a lesser extent, to the type and quantity of instruments used. The model could, therefore, be applied to an existing facility with already defined staff, instrumentation and activities. In this case, the user of the tool can assess both actual performance and possible new staffing needs in the light of a change in the case mix[1] or the acquisition of new technologies. Complex activities, such as the implementation of a cyclotron or of PET–computed tomography (CT) activities for a cancer centre, or the expansion of SPECT myocardial perfusion imaging in the framework of cardiac imaging programmes, involve complex technology and require possibly more staff.

However, this model could also be applied to a planned facility that does not have a consolidated level of activities to refer to. In this scenario, after the evaluation of clinical needs, staffing of the new department could be assessed according to the projected volume of activity, the projected case mix, the modalities to be implemented and the type and quantity of instruments to be used.

1.4. STRUCTURE

This publication describes the methodology adopted to evaluate the staffing needs of nuclear medicine departments at different levels and complexities. For this purpose, the practice of nuclear medicine has been divided into four parts: (1) nuclear medicine imaging, (2) nuclear medicine therapy, (3) radiopharmacy, and (4) cyclotron operations. This methodology was chosen to approach the increasing complexity of practices and their level of organization. The level of complexity is particularly relevant when assessing the staffing needs for radiopharmaceuticals production.

Both imaging and therapy practices could potentially be operational without an in-hospital radiopharmacy and receive radiopharmaceutical preparations from a centralized radiopharmacy outside. Furthermore, if there is no radiopharmaceuticals production from a cyclotron, the in-hospital radiopharmacy may not need an FTE radiopharmacist. If a hospital and a PET radiopharmacy exist within the same institution, staff may be shared (e.g. general radiopharmacists may provide coverage for PET radiopharmaceuticals production).

---

[1] A case mix reflects the diversity, clinical complexity and need for resources of all patients in the department.

The roles and responsibilities of the personnel involved are also discussed in this publication, as well as the role of clinical audits.

2. METHODOLOGY

The tool described in this publication was initially developed as an Excel spreadsheet. The basic parameter for the calculation of staffing needs is the departmental workload, which reflects the level and complexity of clinical activity. The varying complexity and case mix of diverse procedures, specific to each department, may require different levels of staffing.

An on-line version of the IAEA tool to assess staffing needs in nuclear medicine is available on the IAEA's IRIS platform [3]. The tool is also accessible through the Human Health Campus [2]. As well as offering secure data collection and storage, this IRIS form allows complex calculations to be performed on-line. This on-line version of the tool imports the calculation factors from the Excel spreadsheet and contains data validation rules that ensure data integrity. Once the requested information has been provided, the encoded formulas, using the values of the calculation factors, automatically calculate the numbers of nuclear medicine staff required.

Compared with the Excel spreadsheet, the on-line tool has an improved structure and user-friendly interface. Users can navigate intuitively through the multiple tabs on the form to the last page, which provides the calculation results (Fig. 1).

2.1. WORK TIME AND CLINICAL ACTIVITY

The IAEA tool to assess staffing needs in nuclear medicine uses information on the number and type of procedures performed and estimates the FTE of different professionals based on these data. One full FTE is equivalent to 1640 hours of work per year, which is the average work time per year, down from the typical value of 2080 hours/year (40 hours/week × 52 weeks). This reduction considers factors for each staff member such as annual leave, sick leave and absences for training. Using this approach, it is not necessary to make a correction for staff who are not on duty. The ratio between the theoretical working time of 2080 hours and the adopted average of 1640 hours is 1.27, which is frequently used by human resource departments as the replacement factor.

2.1.1. Weight of clinical activities

The most relevant factor for calculating the FTEs is the amount of time required to carry out clinical procedures (i.e. their weight). As an accepted standard, the basic work unit for activities has been set at 15 min. Therefore, the weight of each procedure considers the number of 15 min work units required for each type of procedure, as suggested by the US Centers for Medicare and Medicaid Services [15].

2.1.2. Categorization of procedures and their weights

All procedures, diagnostic and therapeutic, have been categorized into groups according to IAEA Nuclear Medicine Database (NUMDAB) nomenclature [16]. Weights for nurses and physicians in therapeutic procedures have been calculated to reflect typical values of intensity of care and average hospitalization times [17].

Based on the above, and after agreement among the consultants, Table 1 reports the weights related to the total number of work units needed for each staff member for each specific type of procedure. This includes all aspects of the work, from patient admission and interview to discussion with referring

Step #1
Country information

IAEA Tool to assess staffing needs in Nuclear Medicine

Country information

Name of Institution

Type of Institution ○ University-based hospital   Non-University Hospital   Private practice

Street

POBox

City

ZIP

Country

Form completed

Step #2
Equipment and facilities / Radiopharmacy

Equipment and facilities

Planar gamma cameras

SPECT gamma cameras

SPECT-CT gamma cameras

PET scanners

PET-CT scanners

Intraoperative gamma probe

Bone densitometer (DEXA)

Thyroid uptake system

Activity meter

Automatic injector

Ultrasound scanner

Radiopharmacy

For each Radiopharmacy operational Level, select 1 (= yes) when the respective applies, leave 0 (= no), when it doesn't. Higher levels (2 and/or 3) should include the previous level(s).

Supply by Centralized Radiopharmacy (Level 1)

Operational Level 1 is the dispensing of radiopharmaceuticals purchased or supplied in their final form from recognized and/or authorized manufacturers to one or more radiopharmacies. This includes unit dose or multidose doses of prepared radiopharmaceuticals for which no compounding is required.

In-house labeling (up to Level 2)

Operational Level 2 is the preparation of radiopharmaceuticals from prepared and approved reagent kits, generators and radionuclides (closed procedure, with routine use of a radionuclide generator and redissolution of pre-calibrated radiopharmaceutical solutions). Includes level 2b the radiolabelling of autologous blood cells for infection or inflammation imaging.

Complex Radiopharmacy (Level 3)

Operational Level 3 encompasses the compounding of radiopharmaceuticals from ingredients, and radionuclides for diagnostic or therapeutic application requiring open procedures; it also includes the synthesis of positron emission tomography (PET) radiopharmaceuticals, including from generators such as gallium-68/Ga. This level also includes any research and development.

In-house cyclotron

Step #3
Single photon / Therapeutic / PET & PET-CT / Other

Single photon (number of cases per year)

Cardiovascular

Endocrine

Ultrasound studies

Gastrointestinal

Genitourinary

Oncology

Nervous System

Pulmonary

Skeletal

Bone densitometry (DEXA)

Miscellaneous

Subtotal

Therapeutic (number of cases per year)

I-131 Therapy Hyperthyroidism

I-131 Therapy Thyroid Ablation/metastases

Treatment of bone metastases (include 223Ra)

Step #4
Results

Name of Institution

Number of

FTE Nuclear Medicine Physicians

FTE Nuclear Medicine Technologists

FTE Nurses (Diagnostic activities)

FTE Nurses (Therapeutic activities)

FTE Radiopharmacists

FTE Radiopharmacy Technologists

Cyclotron Physicists/Engineers

FTE Administrative staff

Comments

Disclaimer: the above reported results are produced by an automated algorithm: even if care has been used in the tuning of the parameters, they can be considered only as a general guidance on staffing levels. This tool cannot take into account local specificities.

FIG. 1. Steps used in the IAEA tool to assess staffing needs in nuclear medicine.

physicians and attendance at multidisciplinary meetings. Additional aspects, such as administering radiopharmaceuticals and other medication, reporting adverse events and assisting during the acquisition, data processing, storing, archiving and documenting of studies, are also considered.

As an example, consider genitourinary studies, which include both static and dynamic examinations with different characteristics and radiopharmaceuticals. It is assumed that for an average genitourinary scan, a technologist is busy for three work units, a nuclear medicine physician for two and a nurse for one.

In the case of treatments that may require hospitalization (e.g. iodine treatment for thyroid cancer), the weight for staff, particularly for nurses, considers an average hospitalization time of three days. Results from this model could be adjusted to local circumstances if needed.

TABLE 1. WEIGHTS[a] ASSIGNED TO STAFF FOR THE DIFFERENT CLINICAL PROCEDURES[b] BASED ON THEIR COMPLEXITY

Procedure	Nuclear medicine/ attending physician	Nuclear medicine technologist	Nurse
Single photon procedures			
Cardiovascular	5	5	2
Endocrine	3	3	2
Ultrasound	2	0	1
Gastrointestinal	1	3	0
Genitourinary	2	3	1
Oncology	2	3	1
Central nervous system	2	3	0
Pulmonary	2	3	1
Skeletal	3	5	1
Bone densitometry	1	1	1
Miscellaneous	2	3	1
Therapy			
Thyroid benign (I-131)	6	1	4
Thyroid cancer (I-131)	14	2	152
Bone pain palliation (Ra-223; Sm-153; Sr-89; P-32)	6	1	8
Neuroendocrine tumours (I-131-MIBG; Y-90-peptides; Lu-177-peptides)	24	1	152
Radiosynovectomy	2	1	2
Oncohaematology (I-131; Y-90-monoclonal antibodies)	24	1	100

TABLE 1. WEIGHTS[a] ASSIGNED TO STAFF FOR THE DIFFERENT CLINICAL PROCEDURES[b] BASED ON THEIR COMPLEXITY (cont.)

Procedure	Nuclear medicine/ attending physician	Nuclear medicine technologist	Nurse
Prostate cancer (Lu-177-PSMA; Ac-225-PSMA)	24	1	100
Selective internal radiation therapy	10	4	50
PET and PET–CT			
Oncology			
F-18-FDG; F-18-Na; F-18-DOPA	6	6	2
Ga-68-PSMA	6	6	2
Ga-68-DOTA	6	6	2
Others	6	6	2
Cardiology			
Any cardiac procedure	6	6	2
Neurology			
Any central nervous system procedure	6	6	2
Radio guided surgery	2	1	1
Consultations	2	0	2

[a] Weights are expressed as 15 min work units [15]. In the case of treatments of thyroid cancer and neuroendocrine tumours, which require hospitalization and are considered as low intensity of care, work units for nurses have been rounded to 152. This is based on an average in-hospital stay of three days (72 hours × 4 work units/hour) and a ratio of 1.89 (the result of dividing 288/152) between patient beds and number of nurses, as internationally accepted, following a WHO determination [17]. Other types of therapy performed on a day surgery basis follow the same principle.

[b] The procedure groupings are based on IAEA NUMDAB database nomenclature [16].

To ensure a proper interpretation of the results from the model, the following clarifications on staff categories need to be taken into account:

— Physicians of other specialities who are also involved in certain nuclear medicine procedures (i.e. cardiologists, anaesthesiologists or other medical specialists) are not included.
— Nuclear medicine technologists includes professionals who may be designated under alternative names (e.g. radiographers).
— Nurses may include assistant nurses and healthcare assistants (as applies to local regulations). In specific situations, the roles of different staff categories may overlap (see Section 5 on Limitations of the model).
— Staff related to inpatients (ward therapies) may include staff (e.g. nurses) provided from wards outside the nuclear medicine department.
— Ancillary staff groups (e.g. porters and cleaners) are not included.

— For administrative staff, a graded approach has been adopted for calculating the requirements. For this purpose, the model considers a need of 10% of the total staff, plus a proportional fraction related to the increase in workload.

2.2. INFRASTRUCTURE

To ensure a proper estimation of the staffing requirements, it is necessary to take into account not only work time and workload, but also the type of institution (i.e. university based, general hospital, private practice), the type of equipment in use (i.e. SPECT, SPECT–CT, PET–CT, other) and the operational level of the radiopharmacy.

2.2.1. Type of institution

A weight has been given to university based hospitals, reflecting the need for time dedicated to students. The staffing needs according to the clinical workload for university based hospitals is multiplied by a factor of 1.05; for non-university hospitals and private practices, the factor remains equal to 1. The correction factor has been kept at 5% since trainees take up staff time but are themselves also contributing to clinical activity (under the supervision of experienced staff). This correction factor has been applied to all staff categories.

2.2.2. Type of equipment

The level of technology and the number of available equipment items are also considered. In addition to clinical work, staff members need to dedicate time to the following equipment related tasks:

— Preparation of equipment;
— Quality assurance and quality control (QA/QC);
— User-operated maintenance;
— Equipment related administrative tasks.

Time dedicated to equipment related tasks (e.g. QA/QC, calibration, maintenance and the procurement of spare parts or accessories) is expressed as a fraction of FTE staff, based on expert agreement. As an example, Table 2 shows that for a PET–CT scanner, 0.2 FTE of a nuclear medicine technologist is required for equipment related tasks.

2.2.3. Radiopharmacy levels

The model considers radiopharmacy levels as defined in the IAEA QUANUM 3.0 publication [4] and IAEA Operational Guidance on Hospital Radiopharmacy: A Safe and Effective Approach [12], which vary according to the type of facility (e.g. use of generators and/or presence or not of a cyclotron). The model for staffing a radiopharmacy considers that the workload might be shared with technologists, as they are often involved in radiopharmaceutical preparations at operational levels 1 and 2[2]. Depending on the operational level of radiopharmacy, physicians also may have a degree of involvement in managerial responsibilities, contacts with suppliers, preparation of documents for good manufacturing practice compliance, and quality management practices.

---

[2] Many radiopharmacies at operational levels 1 and 2, for in-house use only of radiopharmaceutical preparations, may not have a trained radiopharmacist. In these cases, oversight is provided by the attending physician and/or hospital pharmacist.

TABLE 2. FRACTION OF FULL TIME EQUIVALENT STAFF FOR EQUIPMENT RELATED TASKS

Instrument type	Nuclear medicine physician	Nuclear medicine technologist	Nurse	Radiopharmacist or radiochemist	Cyclotron operator
Planar gamma camera	0.1	0.1	0.1	0	0
SPECT scanner	0.1	0.2	0.01	0	0
SPECT–CT scanner	0.1	0.2	0.01	0	0
PET scanner	0.1	0.2	0.01	0	0
PET–CT scanner	0.1	0.2	0.01	0	0
Intraoperative gamma probe	0.02	0.02	0.01	0	0
Bone densitometer	0.1	0.1	0.05	0	0
Thyroid uptake system	0.05	0.1	0.05	0	0
Activity meter (dose calibrator)	0.01	0.05	0	0	0
Automatic injector	0.01	0.05	0.05	0	0
Ultrasound scanner	0.01	0.01	0.05	0	0
Operational level 1 radiopharmacy	0.05	0.5	0	0	0
Operational level 2 radiopharmacy[a]	0.1	1.5	0	2.2	0
Operational level 3 radiopharmacy[a]	0.1	0.5	0	3	0
Operational level 3 advanced (cyclotron production)[a]	0.1	3	0	0	1

[a]   Staff are in addition to those foreseen for the previous level(s).

2.2.3.1.   Operational level 1

For operational level 1, which involves only dispensing and administering ready-to-use radiopharmaceuticals, 0.5 FTE of a technologist is calculated.

2.2.3.2.   Operational level 2

According to Ref. [12], operational level 2 requires 2 FTE of staff. The model considers these units as technologists or laboratory or pharmaceutical technologists. At the same time, 0.2 FTE of a radiopharmacist or radiochemist is added to include a part-time consultancy on aseptic procedures, microbiological aspects, support in the review of quality documentation and standard operating procedures (SOPs), as well as for training of the other staff. These staff are in addition to those foreseen for the previous level.

2.2.3.3. Operational level 3

At operational level 3, Ref. [12] indicates the need for two radiopharmacists or radiochemists and a 'qualified person'[3]. These three staff members are required to manage production, quality control and final batch release and to provide legal oversight. According to this requirement, the model calculates the need for 3 FTE radiopharmacists, plus a further 0.5 FTE of a technologist. These staff are in addition to those foreseen for the previous levels.

2.2.3.4. Operational level 3 advanced

When operations include a cyclotron, an additional 3 FTE technologists, and 1 FTE of a cyclotron operator are considered. These staff are in addition to those foreseen for the previous levels.

2.3. CLASSIFICATION OF PERSONNEL

The following categories of NMS staff have been considered in the model.

2.3.1. Clinical staff

A member of the clinical staff is a person who is qualified and legally authorized to perform or assist in the performance of a specific professional service. The following staff members are considered to be clinical staff:

— Physicians (nuclear medicine/attending physicians);
— Nurses;
— Nuclear medicine technologists and/or radiographers according to local regulations;
— Radiopharmacists or radiochemists.

2.3.2. Non-clinical staff

The following staff members are considered to be non-clinical staff:

— Administrative staff;
— Cyclotron operators.

3. ROLES AND RESPONSIBILITIES

To account for staff involvement in and time spent on clinical and non-clinical activities, the following responsibilities have been considered for NMS staff for each category in the model.

---

[3] A 'qualified person' is typically a pharmacist, biologist or chemist licensed to oversee the production and release of batches of radiopharmaceuticals for clinical use or for trials according to national rules and EU Directive 2001/20/EC.

3.1. NUCLEAR MEDICINE/ATTENDING PHYSICIANS

The nuclear medicine/attending physician has the following responsibilities [6]:

(a) Interviewing patients;
(b) Defining the clinical appropriateness of and justification for the request or referral, for both diagnostics and therapy;
(c) In accordance with departmental SOPs, giving instructions for the appropriate tests and protocols, keeping in mind the safety of both the patient and staff;
(d) When necessary, tailoring protocols to the needs and condition of the patient;
(e) Interpreting results of diagnostic or therapeutic procedures, based also on clinical information and providing a diagnosis to the extent possible;
(f) Providing training (and education) for technical and junior medical staff;
(g) When in a managerial position, ensuring proper operation of the department and adherence to quality management procedures;
(h) Developing and reviewing departmental SOPs on a regular basis;
(i) Attending multidisciplinary team meetings;
(j) Discussing cases with referring clinicians;
(k) Performing periodic audits of clinical activities;
(l) Reporting adverse events as needed;
(m) Contributing to the departmental quality management system and to internal and external audits.

3.2. RADIOPHARMACISTS OR RADIOCHEMISTS

The radiopharmacist has the following responsibilities [10, 12]:

(a) Overseeing or contributing to the acquisition of radiopharmaceuticals, raw materials and medical devices;
(b) Preparing in a safe and aseptic manner and dispensing radiopharmaceuticals at operational level 3;
(c) Developing and regularly revising departmental SOPs and processing any other quality management recommendations;
(d) Performing QA/QC of radiopharmaceutical preparations and keeping records;
(e) Authorizing the final release of the batch(es) as the 'qualified person';
(f) Reporting adverse events when needed;
(g) Training students and other staff members;
(h) Contributing to the departmental quality management system and to internal and external audits.

3.3. NUCLEAR MEDICINE TECHNOLOGISTS OR RADIOGRAPHERS

The responsibilities of a nuclear medicine technologist typically include some or all of the following:

(a) Preparing the scanner for imaging procedures;
(b) Preparing patients prior to study acquisition;
(c) Administering prepared radiopharmaceuticals (in accordance with local regulations);
(d) Acquiring images;
(e) Processing data;
(f) Displaying images or data;
(g) When needed, preparing in a safe and aseptic manner and dispensing radiopharmaceuticals, and performing quality control;

(h) Measuring the activity of prepared radiopharmaceuticals;
(i) Performing routine quality control of instrumentation;
(j) Contributing to the departmental quality management system and to internal and external audits;
(k) Contributing to the safe handling of radioactive waste.

Technologists are also likely to have additional responsibilities in management (personnel and data), teaching and research. Although in several countries technologists may have very specific duties to perform, the trend is for technologists to take on increasing responsibilities in the management of diagnostic and therapeutic procedures.

3.4. NURSES

The nurse has the following responsibilities [3]:

(a) Booking procedures;
(b) Maintaining medical history records;
(c) Examining vital signs;
(d) Administering drugs and radiopharmaceuticals as prescribed;
(e) Taking blood samples as required;
(f) Explaining procedures to patients and providing support to the receptionist;
(g) Explaining appropriate radiation protection measures to patients and caregivers, especially those comforting children and elderly patients;
(h) Providing immediate support in case of emergencies, particularly for night shifts and weekend work;
(i) Contributing to the departmental quality management system and to internal and external audits.

3.5. ADMINISTRATIVE STAFF

The administrative staff have the following responsibilities [3]:

(a) Liaising with patients and ensuring they are aware of appointment dates and times;
(b) Registering patients and booking all necessary appointments;
(c) Liaising with doctors and senior radiographers or technologists to ensure that bookings are scheduled and prioritized according to SOPs;
(d) Understanding medical terminology to be able to determine the appropriate patient pathway;
(e) Resolving complaints, enquiries and requests from patients and liaising with appropriate members of the multidisciplinary team to deal with these issues;
(f) Organizing and maintaining precise record-keeping, providing details of each patient;
(g) Supervising any change to the processing of booking forms;
(h) Updating and maintaining knowledge of relevant legislation, policies and procedures, and continuously improving skills through appraisal, supervision and training;
(i) Contributing to the departmental quality management system and to internal and external audits.

3.6. MEDICAL PHYSICS STAFFING REQUIREMENTS IN NUCLEAR MEDICINE

Medical physicists play an essential role in modern nuclear medicine and are specifically trained and specialized in this area. They are part of a multidisciplinary team in the nuclear medicine department dedicated to providing safe and effective diagnosis and treatment of disease using radiopharmaceuticals. Staffing needs and roles and responsibilities are described in an IAEA publication [11] to which the reader is advised to refer.

4. CONTRIBUTION TO CLINICAL AUDITS

In addition to their own specific professional roles and responsibilities, all staff need to contribute to clinical audits for their competences and duties, as detailed in several publications. This includes the IAEA publication on the Quality Management Audits in Nuclear Medicine Practices (QUANUM) programme [4] as well as publications on the programme's procedures and positive results [18–20].

4.1. THE QUANUM PROGRAMME: AN AID TO ASSESSING DEPARTMENTAL NEEDS

As health care standards improve globally, providing an optimal service that complies with international levels of quality as well as public expectation requires the implementation of effective quality management systems.

Internal and external audits are an optimal mechanism to evaluate the performance of an NMS. They provide a mechanism for independent assessment of how clinical practice compares with recommended standards, with the aim of identifying areas for improvement or where required standards are not being met (i.e. areas of non-conformance).

The IAEA QUANUM programme described in Ref. [4] provides such a methodology and tools for comprehensive auditing, including all aspects of nuclear medicine practice. Adopting these guidelines will allow an NMS to demonstrate its level of efficiency, quality, safety and reliability in delivering clinical services.

5. LIMITATIONS OF THE MODEL

This publication tries to address staffing needs in terms of FTEs of nuclear medicine physicians, radiopharmacists, administrative staff, cyclotron engineers, nuclear medicine technologists and nurses. The job description of the latter two roles may vary across the world; in some countries, for example, the role of the nuclear medicine technologist is covered by nurses or other types of medical technologist. Conversely, there are situations in which nurses are not employed in nuclear medicine, since technologists are qualified to provide patient assistance and administer radiopharmaceuticals. Therefore, where data produced by this model pertain to the need for both technologists and nurses, the results should be interpreted prudently. As an example: where a technologist also fills the position of a nurse, the staffing needs are the sum of the results obtained for both professional profiles.

Furthermore, this model cannot take into consideration specific local conditions, socioeconomic factors or local regulations. More specifically, this model does not consider the following:

(a) Complex situations such as a department covering multiple work sites;
(b) Details of available equipment and facilities in the radiopharmacy;
(c) Staffing needs related to research;
(d) Staffing needs related to the production of radiopharmaceuticals for commercial purposes or for local non-commercial distribution;
(e) Specific staffing needs related to highly specialized institutions (e.g. paediatric hospitals);
(f) Work processes related to radioimmunoassay.

6. CONCLUSIONS

Clinical nuclear medicine practice requires competent staff to ensure that appropriate, efficient, safe and high quality services are provided. It is essential to define staffing requirements for all professions involved in the practice, not only for new departments but also for those that are expanding their services or introducing new technologies, modalities or applications. Similarly, the introduction of education and training programmes will require adjustments in staffing.

To support nuclear medicine departments in their assessment of staffing needs, the IAEA has developed and made available a tool that calculates workforce requirements through an activity based approach, using inputs that are known or can be easily estimated.

In addition to presenting the model and its related tool, this publication underlines the need for a greater standardization of professional qualifications such as, but not limited to, nuclear medicine technologists and/or radiographers and nurses. The IAEA has issued several publications in this regard.

Finally, it should be considered that the model has been designed to be applied in standard conditions. However, contingencies for emergencies, both health and/or environment related, could also be taken into account.

REFERENCES

[1]   INTERNATIONAL ATOMIC ENERGY AGENCY, IAEA Tool to assess staffing needs in Nuclear Medicine, https://iris.iaea.org/public/survey?cdoc=STFNM001

[2]   INTERNATIONAL ATOMIC ENERGY AGENCY, Human Health Campus, https://humanhealth.iaea.org/hhw/nuclearmedicine/index.html

[3]   INTERNATIONAL ATOMIC ENERGY AGENCY, International Research Integration System (IRIS), https://iris.iaea.org/home

[4]   INTERNATIONAL ATOMIC ENERGY AGENCY, QUANUM 3.0: An Updated Tool for Nuclear Medicine Audits, IAEA Human Health Series No. 33, IAEA, Vienna (2021).

[5]   INTERNATIONAL ATOMIC ENERGY AGENCY, Nuclear Medicine Resources Manual 2020 Edition, IAEA Human Health Series No. 37, IAEA, Vienna (2020).

[6]   INTERNATIONAL ATOMIC ENERGY AGENCY, Training Curriculum for Nuclear Medicine Physicians, IAEA-TECDOC-1883, IAEA, Vienna (2019).

[7]   INTERNATIONAL ATOMIC ENERGY AGENCY, Appropriate Use of FDG-PET for the Management of Cancer Patients, IAEA Human Health Series No. 9, IAEA, Vienna (2010).

[8]   INTERNATIONAL ATOMIC ENERGY AGENCY, Nuclear Cardiology: Guidance on the Implementation of SPECT Myocardial Perfusion Imaging, IAEA Human Health Series No. 23 (Rev. 1), IAEA, Vienna (2016).

[9]   INTERNATIONAL ATOMIC ENERGY AGENCY, Planning a Clinical PET Centre, IAEA Human Health Series No. 11, IAEA, Vienna (2010).

[10]   INTERNATIONAL ATOMIC ENERGY AGENCY, Competency Based Hospital Radiopharmacy Training, IAEA Training Course Series No. 39, IAEA, Vienna (2010).

[11]   INTERNATIONAL ATOMIC ENERGY AGENCY, Medical Physics Staffing Needs in Diagnostic Imaging and Radionuclide Therapy: An Activity Based Approach, IAEA Human Health Reports No. 15, IAEA, Vienna (2018).

[12]   INTERNATIONAL ATOMIC ENERGY AGENCY, Operational Guidance on Hospital Radiopharmacy: A Safe and Effective Approach, IAEA, Vienna (2008).

[13]   INTERNATIONAL ATOMIC ENERGY AGENCY, Leadership and Management for Safety, IAEA Safety Standards Series No. GSR Part 2, IAEA, Vienna (2016).

[14]   EUROPEAN COMMISSION, FOOD AND AGRICULTURE ORGANIZATION OF THE UNITED NATIONS, INTERNATIONAL ATOMIC ENERGY AGENCY, INTERNATIONAL LABOUR ORGANIZATION, OECD NUCLEAR ENERGY AGENCY, PAN AMERICAN HEALTH ORGANIZATION, UNITED NATIONS ENVIRONMENT PROGRAMME, WORLD HEALTH ORGANIZATION, Radiation Protection and Safety of

Radiation Sources: International Basic Safety Standards, IAEA Safety Standards Series No. GSR Part 3, IAEA, Vienna (2014).

[15] US CENTERS FOR MEDICARE AND MEDICAID SERVICES,
https://www.cms.gov/Medicare/Medicare-Fee-for-Service-Payment/PhysicianFeeSched/PFS-Relative-Value-Files

[16] INTERNATIONAL ATOMIC ENERGY AGENCY, Nuclear Medicine Database (NUMDAB),
https://nucmedicine.iaea.org/home

[17] WORLD HEALTH ORGANIZATION, Workload indicators of staffing need (WISN): A manual for implementation, WHO, Geneva (2008),
https://www.who.int/publications/i/item/9789241500197

[18] DONDI, M., et al., Comprehensive auditing in nuclear medicine through the International Atomic Energy Agency Quality Management Audits in Nuclear Medicine (QUANUM) program. Part 1: The QUANUM program and methodology, Semin. Nucl. Med. **47** 6 (2017) 680–686.

[19] DONDI, M., et al., Comprehensive auditing in nuclear medicine through the International Atomic Energy Agency Quality Management Audits in Nuclear Medicine program. Part 2: Analysis of results, Semin. Nucl. Med. **47** 6 (2017) 687–693.

[20] DONDI, M., et al., Implementation of quality systems in nuclear medicine: Why it matters: An outcome analysis (Quality Management Audits in Nuclear Medicine Part III), Semin. Nucl. Med. **48** 3 (2018) 299–306.

ABBREVIATIONS

CT   computed tomography
FTE   full time equivalent
NMS   nuclear medicine service
PET   positron emission tomography
QA/QC   quality assurance/quality control
QUANUM   Quality Management Audits in Nuclear Medicine Practices
SOP   standard operating procedure
SPECT   single photon emission computed tomography

CONTRIBUTORS TO DRAFTING AND REVIEW

Baigorria, S.	Fundación Escuela De Medicina Nuclear, Argentina
Bomanji, J.	University College London Hospital, United Kingdom
Bouyoucef, S.E.	Bab el-Oued Hospital, Algeria
Dondi, M.	International Atomic Energy Agency
Elias, W.E.	Singapore General Hospital, Singapore
Jalilian, A.	International Atomic Energy Agency
Marengo, M.	University of Bologna, Italy
Okolielova, T.	International Atomic Energy Agency
Paez, D.	International Atomic Energy Agency
Pascual, T.	Philippine Nuclear Research Institute, Philippines
Poli, G.L.	International Atomic Energy Agency
Pynda, Y.	International Atomic Energy Agency
Rodríguez Sánchez, D.I.	International Atomic Energy Agency

Consultants Meeting

Vienna, Austria: 18–22 November 2019

ORDERING LOCALLY

IAEA priced publications may be purchased from the sources listed below or from major local booksellers.

Orders for unpriced publications should be made directly to the IAEA. The contact details are given at the end of this list.

NORTH AMERICA

Bernan / Rowman & Littlefield

15250 NBN Way, Blue Ridge Summit, PA 17214, USA
Telephone: +1 800 462 6420 • Fax: +1 800 338 4550

Email: orders@rowman.com • Web site: www.rowman.com/bernan

REST OF WORLD

Please contact your preferred local supplier, or our lead distributor:

Eurospan Group

Gray's Inn House
127 Clerkenwell Road
London EC1R 5DB
United Kingdom

Trade orders and enquiries:

Telephone: +44 (0)176 760 4972 • Fax: +44 (0)176 760 1640
Email: eurospan@turpin-distribution.com

Individual orders:

www.eurospanbookstore.com/iaea

For further information:

Telephone: +44 (0)207 240 0856 • Fax: +44 (0)207 379 0609
Email: info@eurospangroup.com • Web site: www.eurospangroup.com

Orders for both priced and unpriced publications may be addressed directly to:

Marketing and Sales Unit
International Atomic Energy Agency
Vienna International Centre, PO Box 100, 1400 Vienna, Austria
Telephone: +43 1 2600 22529 or 22530 • Fax: +43 1 26007 22529
Email: sales.publications@iaea.org • Web site: www.iaea.org/publications